Voices of Science

Inaugural and Special Lectures published by the University of London Institute of Education

Professor Denis Lawton, *Director of the Institute*
The End of the Secret Garden? A Study in the Politics of the Curriculum
ISBN 0 85473 081 8 — 1979; revised edn. 1982

Professor Brian Holmes, *Department of Comparative Education*
Ivory Towers, the Glass Bead Game and Open Societies: The Social Functions of Comparative Education
ISBN 0 85473 082 6 — 1979

Professor Colin Hindley, *Department of Child Development and Educational Psychology*
Conceptual and Methodological Issues in the Study of Child Development
ISBN 0 85473 097 4 — 1980

Professor Harold Rosen, *Department of English as a Mother Tongue*
Neither Bleak House Nor Liberty Hall: English in the Curriculum
ISBN 0 85473 109 1 — 1981

Professor Malcolm Skilbeck, *Department of Curriculum Studies*
A Core Curriculum for the Common School
ISBN 0 85473 210 1 — 1982; revised edn. 1985

Professor Keith Swanwick, *Department of Music Education*
The Arts in Education: Dreaming or Wide Awake?
ISBN 0 85473 147 4 — 1983

Professor Mark Blaug, *Department of Economic, Administrative and Policy Studies in Education*
Where Are We Now in the Economics of Education?
ISBN 0 85473 159 8 — 1983

Professor Peter Gordon, *Department of History and Humanities*
Purpose and Planning in the Humanities Curriculum
ISBN 0 85473 175 X — 1984

Professor Hazel Francis, *Department of Child Development and Educational Psychology*
Minds of Their Own: Recent Trends in Educational Psychology
ISBN 0 85473 194 6 — 1984

Professor Norman Graves, *Pro-Director, Professional Studies*
Teacher Education in Adversity
ISBN 0 85473 223 3 — 1985

Professor Jon Ogborn, *Department of Science Education*
Voices of Science
ISBN 0 85473 234 9 — 1985

Professor Celia Hoyles, *Department of Mathematics, Statistics and Computing*
Culture and Computers in the Mathematics Classroom
ISBN 0 85473 235 7 — 1985

Jon Ogborn
Professor of Science Education

Voices of Science

An Inaugural Lecture delivered at the
University of London Institute of Education on
Wednesday, 5 June, 1985

Institute of Education, University of London
Distributed by Turnaround Distribution Ltd.

First published in 1985 by the University of London Institute of Education,
20 Bedford Way, London WC1H 0AL

Distributed by Turnaround Distribution Ltd., 27 Horsell Road, London N5 1XL
(telephone 01-609 7836)

Cover photography by Caroline Rees; cover design by Joan Rose

ISBN 0 85473 234 9

British Library Cataloguing in Publication Data
Ogborn, Jon
Voices of science : an inaugural lecture delivered at the University of London Institute of Education on Wednesday, 5th June, 1985.
1. Science — Philosophy
I. Title II. University of London.
Institute of Education
501 Q175

ISBN 0-85473-234-9

Printed in Great Britain by Reprographic Services (typesetting by Joan Rose)
University of London Institute of Education

113/I1-001/000-0885

Prefatory Note

The Inaugural Lecture, as befits a lecture by a scientist, included a number of demonstrations. Since their main purpose was to illustrate a thesis about realism, to attempt to include them as drawings or photographs would hardly be to the point. I hope that what I intended to convey by them will still emerge, even in their absence. I am grateful to David Holland for his assistance with the demonstrations.

J.M.O.

Acknowledgement

The author and the Institute are grateful to Faber and Faber Publishers for permission to reproduce 'Ignorance' from *The Whitsun Weddings* by Philip Larkin.

Voices of Science

Those of us who practise and teach science confront today a number of conflicting and paradoxical views of the nature of our pursuit. It is not too much to say that science is under radical attack, and that many, even within science education, have lost confidence in the value of what science can offer. This has arisen in part through confronting the problems of what it would be to teach science to the whole population, and in part from philosophical and sociological critiques of science itself. Since I believe that a large part of this self-doubt is based on a wrong understanding of the kind of knowing that is science, my purpose in this lecture is to sketch out a personal position about the nature of science, and to say something about what that position means for the problems of teaching science.

I want to begin with a remarkable and too little noticed essay of Michael Oakeshott's, 'The Voice of Poetry in the Conversation of Mankind'. The echo of his title in my own is intended. That essay speaks to me personally with great force. In it, I think, are some very deep insights about the nature of knowing; and, as I have said, it is doubts about the kind of knowing that is science which I think lie behind many present concerns and anxieties in science education.

Oakeshott's central image is that of a conversation; of exchanges which hang together without having a final goal, which react to one another without having fixed or external criteria of consistency; a conversation as an activity which is an end in itself driven by various appetities — whether appetites to understand, to be playful, or to master events. This image Oakeshott intends as his image for human knowing:

> As civilized human beings, we are the inheritors, neither of an inquiry about ourselves and the world, nor of an accumulating body of information, but of a conversation, begun in the primeval forests and extended and made

> more articulate in the course of centuries . . . Of course there is argument and inquiry and information, but wherever they are profitable they are to be recognized as passages in this conversation, and perhaps not the most captivating.[1]

What Oakeshott does is to rest everything on human activity, and to draw out the full consequences of this view, as few others save especially Wittgenstein and, in science, Polanyi or, in psychology, Piaget, have been able to do. For him there is no rationality not made by us, no standards of truth not won by human effort, and no objectivity or reality not made objective and real by our own strivings. The very nature of what we call 'truth', 'objectivity', 'reality', and 'rationality' belong *in* the conversation, not outside it.

Oakeshott writes of 'voices' in this conversation, one but only one of which is that mode of conversing we call science:

> Each voice is the reflection of a human activity, begun without premonition of where it would lead, but acquiring for itself in the course of the engagement a specific character and a manner of speaking of its own . . . This conversation is not only the greatest but also the most hardly sustained of all the accomplishments of mankind.[2]

Let me now be very clear. In my understanding, it does *not* follow from these thoughts that 'truth', 'objectivity', 'reality', and 'rationality' are illusory. It does *not* follow that science has no business to speak of facts discovered or established, of theories explaining parts of reality, of searching for truth. It *does* follow that science is not a neutral voice, free of idiom, outside or above the human conversation, with special access to some already given, uninvented way of thinking. But again, from this last, it does *not* follow, as one might imagine that it should, that the voice of science lacks special qualities, and in particular that it lacks qualities to do with objectivity, reality and truth.

This seems to be a paradox. What I have just said flatly contradicts what has recently seemed to some in science education (and elsewhere) to be an obvious consequence: that if science is simply a fallible human fabrication then any pretensions to have 'got it right' are at best self deception, at worst self-aggrandisement. But to suppose this is to misunderstand the very point. To argue in that way is to appeal to notions of 'objectivity', 'reality' and 'truth' which are imagined as given, as not made by us, when the point is that this is an impossible fantasy. It is to reproach science for not doing the impossible. We had better argue that, fantasies of ultimate certainty apart, the remarkable quality of the voice of science in the human conversation is that it has, astonishingly, given such qualities a real and reliable meaning for us. Put simply, science has profoundly altered what it is we call 'being rational', not by going beyond the human sphere but by endeavour within it.

Let us try, speculatively, to look at this in evolutionary terms. Every organism, to survive, has we may suppose to deal with what the future will bring. The simplest amoebae, so far as we can tell, react only to that immediate future we call the present, moving in response to concentration gradients of nutrient, light and so on. Insects seem to have one or more stereotyped behaviour patterns, whose potential may be genetically coded, which look to us like purposive responses to what the future has brought, such as the approach of an enemy or of a meal. To deal thus with the future requires the processing of information, and it seems likely that central nervous systems are usefully regarded as a good evolutionary strategy. But a nervous system can go beyond taking in information and generating reflex actions; it can loop internally and feed processed information back to itself for further work. This, ultimately, would be what is needed for a nervous system to deal with the future by imagining it, though we have little idea how the trick is done. So, 'in our minds', as an important survival strategy, we make futures and anticipate events, enabling us to manage what the actual future brings rather better. It is not clear how far this strategy extends to other organisms; it may be that the imagining of absent things is uniquely human.

We cannot make real futures, in all their complexity. What we can do is to make 'model' futures, more or less simplified and idealized. This may be a good device, to make it possible to think fast enough to anticipate the future before it arrives. It would be no wonder if humanity, with these gifts, had imagined the idea of knowing for certain, and had developed a considerable passion for this notion of being absolutely sure once and forever.

Further, like many other forms of life, we have evolved strongly social survival strategies. We survive together, and develop our models of possible futures not alone but mutually. In particular, our cognition is both social and individual, as Pask[3] so clearly recognizes. Thus it may be that the social construction of reality could perhaps be understood. Wittgenstein knew it well:

> 'We are quite sure of it' does not mean that every single person is certain of it, but that we belong to a community which is bound together by science and education.[4]

Wittgenstein also saw clearly that, as I argued previously, human rationality is what it makes of itself:

> Is it wrong for me to be guided in my actions by the propositions of physics? Am I to say I have no good ground for doing so? Isn't this precisely what we *call* a 'good ground'?[5]

Science, then, as I understand it, is the mutual and consensual human construction of organized and related images of those aspects of the world which we can understand as being as they are and not as we might wish or prefer them to be. What is mutual is, however, always in danger of being tainted by collusion: for it to be less so what is needed is in addition an independent individual critical passion. Sometimes this critical passion works on fine detail, noticing something amiss. Sometimes it reshapes the whole conversation, proposing completely new ways of speaking. Somewhere here, I think, are the sources of our puzzlement as to whether truth is to be identified with impersonality or with personal commitment; a puzzlement which led Polanyi, who knew that science needs both, to write:

> The decisive reason why . . . obviously inadequate formulations of the principles of science were accepted by men of great intellectual distinction lies in a desperate craving to represent scientific knowledge as impersonal.[6]

A similar idea is expressed by Oakeshott, writing about the teacher:

> . . . like many of us, he may be expected to have a superstitious prejudice against the human race and to be satisfied only when he can feel himself anchored to something for which human beings are not responsible. But he must have the courage of his circumstances. This man-made inheritance contains everything to which value may be attributed; it is the ground and context of every judgement of better and worse.[7]

* * *

I turn now to something of the origins of the conversational voice of science. As we know it today it is only a few hundred years old, and seems to have depended on a certain social order for its growth. One who began the discussion as we know it today was Francis Bacon. Few appear to read him, though many refer to him. Those who read about him read of a primitive who recommended a misguided process called 'induction'.

However, let Francis Bacon speak for himself:

> Those who have handled sciences (which in Bacon's time meant any organized knowledge; the usage restricting 'science' to natural science grew only in the nineteenth century) have been either men of experiment or men of dogmas. The men of experiment are like the ant, they only collect and use; the reasoners resemble spiders, who make cobwebs out of their own substance. But the bee takes a middle course: it gathers its material from the flowers of the garden and of the field, but transforms and digests it by a power of its own . . . Therefore from a closer and purer league between these two faculties, the experimental and the rational (such as has never yet been made), much may be hoped.[8]

'Baconian Induction' is a philosophical myth, in my reading of Bacon. He writes explicitly against it: '. . . induction which proceeds by simple enumeration is childish; its conclusions are precarious and exposed to peril from a contrary instance'.[9] And again: '. . . to conclude upon a bare enumeration of particulars (as the logicians do) without instance contradictory, is a vicious conclusion'.[10]

What Bacon was against was not so much scholastic appeal to authority, as the posing as evident of very general principles drawn from a few facts seen in a certain light, and then — what he truly detested — the 'deduction' from such principles of 'facts' about how things are. Bacon thought one should work a lot harder: he was particularly scathing about the 'vain and hurtful . . . conceit . . . that the dignity of the human mind is impaired by long and close intercourse with experiments and particulars'.[11]

It is vital to recall that when Bacon wrote, the essential league between the experimental and the rational had indeed 'never yet been made'. Bacon wrote his conversational contribution 'without premonition of where it would lead', in Oakeshott's phrase. It is true, as records of the early Royal Society clearly show, that much immediately post Baconian science was ant-like. Only after Newton, Lagrange, Laplace and others would it be possible to discern how powerfully bee-science could combine ant and spider.

Why should so tough-minded and practical a man as Francis Bacon have fallen into such undeserved disrepute? One reason, I suspect, is the astonishing success of his programme. Science became the envy of thought, admired and imitated more than a single voice in a conversation ought to be. The false question of how science seemed to have managed the impossible feat of arriving at sure and certain knowledge agonized thinkers for a couple of centuries. Bacon's answers to the different question of how to get ahead did not answer this false question.

Instead, in looking for a solid basis for the impossible ideal of certainty, philosophers stressed the very peculiar idea (against which Bacon argued strongly) that first-hand 'sense impressions' have a special primacy. That idea gained a sway which is still with us today. For Bacon, however, '. . . the secrets of nature reveal themselves much more readily under the vexations of art than when they go their own way'.[12] Science had to learn the art of vexing nature, which we call experiment. I shall argue in a moment that experiment is the means by which entities in science are constructed as real.

Secondly, thinkers became obsessed with scientific theories, to the exclusion of scientific practice. Having failed to answer the impossible question of how theories could be known to be certainly true, the wheel of discussion has turned full circle so that we are invited to agree that no fact is secure either, being — as the phrase has it — 'theory loaded'. Ian Hacking gives a good account of the more sensible meaning given to this phrase by its originator, N R Hanson.[13, 14]

All this, in addition to the philosophical headaches it has caused, has also led to attacks on school science as being 'naïvely Baconian'. For my part, I wish it were truer to Bacon than it is.

* * *

So far we have been thinking *about* science. But science ultimately thinks about things rather than about thoughts, which is why there is a certain amount of hardware here with me. So let us now turn to that hardware, assured by Bacon that doing so will not impair the dignity of our minds, to illustrate some of the twists and turns of his league between ant and spider; between the experimental and the rational. In doing so, I shall try to show how 'reality' is constructed in the activity of science, so that for people like me atoms or genes are as real as are sticks and stones and people and animals.

First, here we have a length of copper wire. If I pull it a little it behaves like a spring: it comes out and goes back again. How can that be? Does some space open up in it? If I pull a little more, it stretches but does not now go back. It stretches by a very surprising amount before it breaks. Where it broke, the metal has flowed to form a neck. Again, how can these things be? One response is ant-like: to measure carefully — does it get thinner?; to try other substances — are they the same? (they are not all the same); to record, tabulate and use. In that way bridges and buildings and, I have to say, guns, were made before anyone had the slightest idea why metals can behave like this. An understanding had to wait, first on the accidental discovery of X-rays which revealed the very symmetric arrangement of atoms — a spider-idea of ancient origin — in metals such as copper, and second on careful bee-inspired calculations which showed that the layers of atoms simply would *not* slip under gentle pulls such as I made and on the new idea that slipping might be due to small faults in that perfect structure, called dislocations, running through the layers of atoms. In this way we came to understand, in our century, the reasons behind millenia-old crafts of sword-making and metal shaping.

This I think is what Bacon meant when he proposed to 'analyse nature by proper rejections and exclusions'.[15] So far as pulling and shaping metals is concerned, it is difficult to see any 'theory loaded observation' here, I think. And science is full of such texturally complex prodding of possibilities. Another example is that of Herschel noticing that his hand felt warmer behind some of his coloured telescope filters than it did behind others, which led him to the idea of infra-red radiation.

But notice what happens in addition. A model is introduced, in this case via investigations into atoms having nothing whatever to do with metals and their structure, indeed being part of a major movement in scientific ideas towards an atomic world picture. At the beginning of the century, Mach, Ostwald and others denounced atoms as metaphysical constructs; a generation later dislocations were proposed as real faults in patterns of real atoms. It still had to be twenty years more before anyone 'saw' them. As one who had spent a working life predicting the behaviour of dislocations he had never had direct evidence for, Charles Frank said, 'When we could see them, it ceased to be a purely deductive science. It turned out that we had identified all the pieces on the chess-board, and the rules by which they moved; but the games they played required analysis continuing *ad infinitum,* or at any rate up to the present day.'[16]

So, given a model, we look for ways in which it may be right and ways in which it may be wrong. All suggest further things to try. As we proceed, we become realists about what we are looking into, through the process of treating it as real, that is of manipulating and prodding and then using 'it'. Soon the ideas of atoms, of regular structures, of faults, become real; first by being used to explain and then by being built in to explanations of other things. Reality arrives most strongly when we actually *exploit* our ideas, for example so as to deliberately make stronger materials to a prior specification. I am much indebted to the writings of Ian Hacking for this essential insight.[17]

The previous example does little violence to common sense, but the next — a ball thrown in the air — does. Common sense can explain it easily enough — my throw gives it a force which carries it up until that force is used up, and the ball then falls as it would if I let it go from a height. It took the genius of Galileo and Newton to see that this natural explanation could be replaced with profit by a much more unnatural one.

Rather than write the theory for you, I shall show it with a 'theory engine' — the computer. The Newtonian idea is that forces just change motion, and nothing else. So if the computer 'ball' is set moving, it just goes on moving if I don't hit it again. If I do hit it, it changes its motion, and again moves continually. Newton's first idea, taken from Galileo, was that *moving needs no explanation*. It sounds crazy, but there is worse to come. The real ball thrown in the air rises and then falls, and travels along a rather nice curve. Newton's other idea, which he thought of to explain the solar system, not apples falling, was that there isn't any upward force carrying the ball up at all; that the *only* force is a *downward* one, made by the earth pulling on the ball. You see on the computer, that if I push downwards, and only downwards, all the time, the 'ball' does just what the real one does. The path produced is a parabola, and this follows rather easily from the details of how forces change velocities. The parabola is what we get by combining a downward acceleration with a steady sideways drift — that is, a sideways motion with *no* force. And the ball is, when I throw it for you, a good enough mathematician to follow the curve.

Out of all this, an unobservable entity, 'gravity', began its real existence for scientists. It is still not however fully real, for we cannot manipulate it in a thing-like way. That may yet come: people think they know how to send waves through gravity, but they cannot yet detect such waves, if they exist. You will notice that this voice of science is far from a positivist voice. It is realist, but realist about things, like gravity, which most people would think to be as unreal as anything could be. And out of all this came the crazy, but true and for humanity deeply significant notion, that we are sailing through space stuck to a rocky sphere. Thus can science change our idea of what we are.

Lastly, let me illustrate further the construction of new realities through experimental manipulation.

Here is a coil of wire. Connected to a supply of electricity it affects a magnet, though there is nothing otherwise magnetic about the coil. A multitude of tiny magnets shows that the effect has a definite shape, and exists everywhere around the coil. One could still think of this 'field', as it is called, as unreal; as just another way of saying that magnets line up. But Faraday found, after frequent failures, that changes in the field could make electricity. So I can have a coil here, and another not connected to it at all, and make a lamp in the disconnected coil light up. Or I can make a metal ring jump in the air. Such phenomena, when people had learned to create them at will, led in two totally contrasting directions. One was that of use and exploitation, through which the abstract 'fields' became constructed as real, becoming parts of electric motors and dynamos, and thus of underground trains and food mixers.

The other was to construe them as real in a much more daring and theoretical sense: Faraday, and then Maxwell, had the idea that such a field might move, might travel on its own through space. Yet, as Feynman has put it, this was to ascribe reality to mere numbers attached to points in space. And this idea came both to be an explanation of what light is — that is, electricity and magnetism without wires — and to suggest how to make such waves in a field, giving us radio and television. The television news may or may not be true, but we all suppose the signals carrying it to be real.

The point is not the technology, important though that is. The point is that the intricate to and fro in the conversation between the voices of theory and experiment makes new realities. In the process, human rational capacities are extended, and we get new concepts and new ways to think.

* * *

I want to conclude with an agenda of issues and questions which deserve reflection in science education; an agenda for which what I have said so far is no more than a necessary preliminary. It is only an agenda; I shall do no more than indicate points of contact with what I have said so far.

First, science and culture. It is clear that science is an offshoot of Western industrialized society, though not merely or only that. It shares, and may exaggerate, features of that culture, including racism and sexism as well as the Protestant work ethic and a tension between individualism and collectivism. None of that could easily be otherwise. But, by concerning itself with what can be understood as independent of human wishes, science is necessarily involved in attempting to identify what might be consensual across cultures. The voice of science is one committed to the attempt to extend the areas of discourse which can be seen as neutral and objective. In that attempt it has had considerable success.

Once we clearly understand that 'neutrality' and 'objectivity' are human constructions built in part through science, not wholly given prior to science, we may perhaps abandon as immature the two competing claims there are about these qualities. I mean the exaggerated scientistic claims for science possessing them as absolute qualities, and the no less exaggerated counter-claims that science has no more of such qualities than any other discourse. I think that it is John Ziman who has, amongst recent authors, written best about the sense in which scientific knowledge is truly reliable, in any humanly reasonable sense of that term.[18] And I think that we may quite properly value this product of our culture.

Next, the exclusive voice of science. An important difficulty with science is that, following Bacon, it avoids knock-down arguments. Trails are followed in immense and seemingly tedious detail, not obsessively but because fatal flaws are often well hidden. Thus its arguments are technical and complex, and its discourse is fully available only to those who have survived following all the twists and turns of such argument. Further, the attempt at neutrality and objectivity has led to much use of formal argument, particularly of mathematics. Science, in a word, is not invariably conversable. But it can be made more so, as great popularizers have shown, and an essential question for science education is to consider its popularizing role. If it avoids such a role, it may offer only the substance, and not the importance, of scientific thought.

Now for a related matter — the rejection of science. Science is currently rejected for two contradictory reasons; that it is not necessarily true, or that it is all too horribly true. The first is a reaction to past, and present, scientistic overvaluations of science and its methods. Not long ago, the science teachers' journals frequently contained pleas to 'be more scientific' in our approach to teaching; now they are more prone to carry views of science as at best a temporary gloss on experience, with teachers concerned that they, in the name of offering an understanding of the natural world, are forcing as dogma on the young what is no more than a frail and possibly self-serving set of passing schemes of interpretation. At its most extreme and foolish this produces proposals for 'content-free science' (unreliable methods without unreliable results) and for 'children's science' (confusing respect for the personal validity of a child's thoughts with that other respect which requires one to teach the child something).

The second is the wider rejection of science as having delivered, not material comforts and helpful artefacts, but danger and destruction; not beneficent knowledge but knowledge we would do better not to have. A part of that knowledge is our understanding of sources of energy. First came knowledge of how to exploit the energy of coal and oil, which as we understand it, is to tap the ancient energy of the sun, getting access to temperatures of a few thousand degrees. Next it was understood that our earth is the cold ashes of even more ancient exploding stars, so that in using uranium we are mining back into their deep interiors, reaching now to temperatures of millions of degrees. Finally, in fusing hydrogen to helium, we are tapping sparks from the first few minutes of the exploding universe, when they were made, when temperatures were in the region of billions of degrees. It could have been that we live in a less awesomely potent universe, but it seems that we do not.

Neither reaction is without sense, given the excessive trust that has been too often placed in just one voice in our human conversation. But it seems to me that there is something almost Freudian in this wish to murder the god that is felt to have failed.

Consider next the dogmatic voice of science. Those who converse within science find its ideas compelling — for no better reason than that the rules of the discourse exclude ideas which are not compelling. This sounds like a recipe for conformity, but it is actually a recipe for revolution. If things absolutely must fit, then when they don't, they must change. But the compulsive power of that which all fits together does lead to science being experienced as dogma. The difficulty is well put by Polanyi:

> . . . a frame of mind in which I may hold firmly to what I believe to be true, even though I know that it may conceivably be false.[19]

Somehow, science teaching has to capture a way of speaking within such a frame of mind.

Finally, I want to mention the imaginative voice of science. We earlier considered the throwing of a ball. Newton's achievement was not so much to explain that, but rather to imagine away the earth, so as in imagination to bring it back and make the moon go round it because of its gravitational pull. Humanity has always lived with gravity: Newton understood it by imagining being without it. In our own times, further imaginings are afoot. 'Obviously' an object which moves just goes from place to place, keeping its identity. But is it so obvious? How do we explain existence, identity, itself? That is what recent physics is up to. To do that one must imagine the properties of nothingness, and indeed recent thought sees existence as a spark bursting spontaneously from the very structure of nothing at all.

Galileo understood very well the paradoxical nature of the league of reason and experiment, in which attention to how things are may lead to the strangest of thoughts about why they might be so; that science is *not* 'organized common sense':

> Reason hath committed such a rape upon the senses as, in despite thereof, to make her mistress of their credulity.[20]

This then is what we must understand if we want to understand science: how it is that the rational and experimental modes of its voice converse together, and in so doing create models of the world which are as real and reliable as any that exist, whilst at the same time being amongst the most imaginative of human constructions. And we have to understand that it is in the concrete doing of science that this reality is made.

* * *

Science is one voice conversing with others in an uncertain world; its discourse seeking to understand what will be whether we wish it or no. And in this sense the voice of science converses with the deepest urges of humanity:

> Strange to know nothing, never to be sure
> Of what is true or right or real,
> But forced to qualify *or so I feel,*
> Or *Well, it does seem so:*
> *Someone must know.*
>
> Strange to be ignorant of the way things work:
> Their skill at finding what they need,
> Their sense of shape, and punctual spread of seed,
> And willingness to change;
> Yes, it is strange,
>
> Even to wear such knowledge — for our flesh
> Surrounds us with its own decisions —
> And yet spend all our life on imprecisions,
> That when we start to die
> have no idea why.

Philip Larkin, 'Ignorance' [21]

References

1. Michael Oakeshott, 'The voice of poetry in the conversation of mankind', in *Rationalism in Politics,* p.199. Methuen: London 1962.

2. *ibid.* pp.199-200.

3. Gordon Pask, *Conversation Theory: applications in education and epistemology*. Elsevier: Amsterdam 1976.

4. Ludwig Wittgenstein, *On Certainty,* paragraph 298. Edited by G E M Anscombe and G H von Wright, translated by Denis Paul and G E M Anscombe. Basil Blackwell: Oxford 1977.

5. *ibid.* paragraph 608.

6. Michael Polanyi, *Personal Knowledge,* p.168. Routledge and Kegan Paul: London 1958.

7. Michael Oakeshott, 'Learning and teaching', in R S Peters (editor) *The Concept of Education,* p.162. Routledge and Kegan Paul: London 1967.

8. Francis Bacon, *The New Organon,* paragraph XCV. Bobbs-Merrill: Indianapolis 1960.

9. *ibid.* paragraph CV.

10. Ian Hacking, *Representing and Intervening,* p.247. Cambridge University Press: Cambridge 1983.

11. Francis Bacon, *op.cit.,* paragraph LXXXIII.

12. *ibid.* paragraph XCVIII.

13. Ian Hacking, *op.cit.*

14. N R Hanson, *Patterns of Discovery*. Cambridge University Press: Cambridge 1959.

15. Francis Bacon, *op.cit.*, paragraph CV.

16. Charles Frank, *Physics Bulletin* Vol 36, No 4, April 1985, p.154. Institute of Physics: London 1985.

17. Ian Hacking, *op.cit.*

18. John Ziman, *Reliable Knowledge*. Cambridge University Press: Cambridge 1978.

19. Michael Polanyi, *op.cit.* p.214.

20. I am indebted to Professor P J Black for this remark of Galileo.

21. Philip Larkin, 'Ignorance', in *The Whitsun Weddings*. Faber and Faber: London 1964.